WRITING A
SCIENTIFIC RESEARCH PAPER
THE QUICK AND EASY WAY

WRITING A
SCIENTIFIC RESEARCH PAPER
THE QUICK AND EASY WAY

DR. ARIE VAN DER MEIJDEN

TABLE OF CONTENTS

PREFACE

Since you are reading this book, you probably need to write a scientific research article, but you don't know how to start. Or maybe you got stuck with your article and didn't know how to continue. Perhaps you are just looking for tips to write a better scientific article in less time? In any of these cases, this book is for you.

At the beginning of my career as a scientific researcher, I usually got stuck when writing my articles and wasted a lot of time. The task seemed overwhelming, and I did not know what to do next. But over the years, I have developed a method for writing articles that is much more effortless and helps me avoid getting stuck. As a research biologist, I have now written over 50 articles. I have helped many Master's and Ph.D. students overcome the many mistakes all beginners make when writing a research article. Now, I want to share that experience and help you to speed up *your* writing process.

Much of what this book will cover is general to article writing in most scientific fields - biology, biomedical sciences, astronomy, geology, computer science, etc. If you have results to report, the method and tips in this book will help you to avoid getting stuck and save you much time.

You will:

- learn about the different types of scientific articles

- understand the different parts that make up a research article
- use my method to write each section in the most effortless order.
- learn to avoid several common mistakes
- get tips on how to collaborate with your co-authors, how to select a journal, and
- get a brief overview of the submission and review process

This book aims to help people who are relatively new to writing scientific research articles become more effective at writing. You will be able to start writing immediately, *even while reading this book.*

In this book, I will assume you have already started doing your research and that you know the literature in your field – I cannot help you with the science. This book also assumes you know how science is written up, perhaps from the articles you read. If you do, this book will help you structure your research article writing in a goal-oriented way and not waste time and energy on beginner mistakes.

So if all that sounds good to you, let's dive into it!

1. THE DIFFERENT TYPES OF SCIENTIFIC ARTICLES

Here you are. You have done your research. You may have done some experiments, and you have analyzed your data. Now it's time to write it up. But you're unsure of where to start?

Before writing a scientific article, it is vital to understand how they are built up. Knowing which parts go where will not only help you write your own articles, it also greatly speeds up reading articles; You will know exactly where to look for that certain titbit of information you need. But they are all organized in different ways; there are many kinds of articles, each with its own structure. For instance, there are original research articles, reviews, commentaries, case reports, method notes, concept reports, clinical suggestions, etc. Each of these article types has a different function and structure to go with it. Although there are many types of articles (and some scientific fields have their own particular types), here I will separate them into three major types: Review Articles, Comment/Discussion Articles, and Research Articles.

REVIEW ARTICLES

These give an overview of the literature on a specific topic or field. Essentially, in a review paper, an expert takes you by the hand and tells you the most important things in their field from their point of view. So always look for review papers when you are getting into a new subject- they will give you an overview of the literature and save you much time that you would have otherwise spent to find and interpret these papers on your own. Sometimes they contain the essence of a whole career in a particular field. For example, we recently wrote a review paper summarizing 341 articles from several decades of literature. We wrote it to help our colleagues find information scattered across the literature and to relate it all in one place. This way, we hoped to stimulate other researchers to look into this topic. Although we will mainly focus on Research Papers, much of the methodology and tips in this book can also help write a review paper.

Examples of review articles:

- Simone, Y, Van der Meijden, A. (2021) Armed stem to stinger: a review of the ecological roles of scorpion weapons. J Venom Anim Toxins incl Trop Dis. 2021;27 https://doi.org/10.1590/1678-9199-JVATITD-2021-0002
- Delshadi, R. et al. (2021) Development of nanoparticle-delivery systems for antiviral agents: A review. Journal of Controlled Release 331: 30-44. https://doi.org/10.1016/j.jconrel.2021.01.017
- Singh, J., and Singh, S. (2022) A review on Machine learning aspect in physics and mechanics of glasses. Materials Science and Engineering: B 284: 115858 https://doi.org/10.1016/j.mseb.2022.115858

COMMENTARY ARTICLES

Some articles are meant as comments, opinions or are part of discussions in the scientific literature. These are structured more like letters and tend to be very short. They convey a point of view or a disagreement about scientific topics between different authors. Apart from addressing a specific issue, their structure varies by journal and field. I will discuss this kind of article here only.

Examples of commentary articles:

- Wenz, S.E. (2019), What Quantile Regression Does and Doesn't Do: A Commentary on Petscher and Logan (2014). Child Dev, 90: 1442-1452. https://doi.org/10.1111/cdev.13141
- Dominy N.J. and Harris J.M. (2022) Adaptive optics in the Arctic? A commentary on Fosbury and Jeffery (2022). Proc. R. Soc. B.2892022152820221528 https://doi.org/10.1098/rspb.2022.1528

RESEARCH ARTICLES

The most common type of scientific article is the Research Article. The goal of the research article is to convey the results of a research project in a standardized format: Abstract, Introduction, Methods, Results, and Discussion. This standardized format is necessary to help readers quickly find the parts they are interested in; Perhaps some readers want to know what method you used, so they can go straight to the methods section; Some readers want the gist of the paper, so they will go to the abstract. More importantly, the Results and Discussion sections separate objective results from interpretation and speculation. This type of paper will be discussed extensively in this book.

Each journal may have a slightly different way of parsing a research paper. Some have additional sections, such as a "Conclusions" section after the Discussion or a "Highlights" section to go with the Abstract. Some journals join the Results and Discussion sections into one section. In the following chapters, we will discuss a research article's parts.

There are also a few journals that mix things up more than usual. These are typically high-impact journals with space constraints because they need to fit everything in a limited number of printed pages. An example is the journal Nature. But even there, they still use most of the standard elements of a research article[1]:

- Abstract
- Introduction
- Methods
- Results
- Discussion

---

[1] I use the words "article" and "paper" interchangeably throughout this book. That is a bit wrong, of course, as a scientific article is just a scientific paper that has been published, so the concepts overlap but are a bit different. Just know that in this course, I use them interchangeably. Also, it is good to know that it is usually called a manuscript until your article is accepted for publication.

2. ARTICLE ANATOMY

2A. THE TITLE PAGE

As you learned in the previous chapter, a scientific research article is made in a standardized format with different sections, each with another function and content. When writing an article, it is important to understand what goes where and why these sections are separate. Let's go through the different research article sections together in the order you read them. But remember, this is NOT the order in which you will write your article! We will get into how to write it in the following chapters.

THE TITLE

The title is the most important part of the article, as it is the first thing a reader will read. The title acts as a label of what your article is about, and aids your readers in deciding whether to read it or not. Since hundreds of articles come out monthly in each scientific field, nobody can read everything. So a good descriptive title helps your reader quickly scan the literature and select the articles that interest them.

Most printed journals also have something called a "running title." This is a shorter version of the title that appears at the top of each printed page or every other page. This is so that when thumbing through a journal, you can quickly identify where one article ends, and a new one begins. It is primarily helpful in printed journals and less in digital formats.

THE AUTHOR LIST

Usually, anybody who has contributed significantly to the article is on the author list. What constitutes a "significant contribution" may differ between institutes and fields. Usually, everyone who has done data collection, analysis, and/or written the manuscript is in there, but there can also be people who helped by, for instance, supplying financing or research infrastructure. Who is an author on the paper is usually decided by one or both of the two most important authors of an article: the first author and the communicating author.

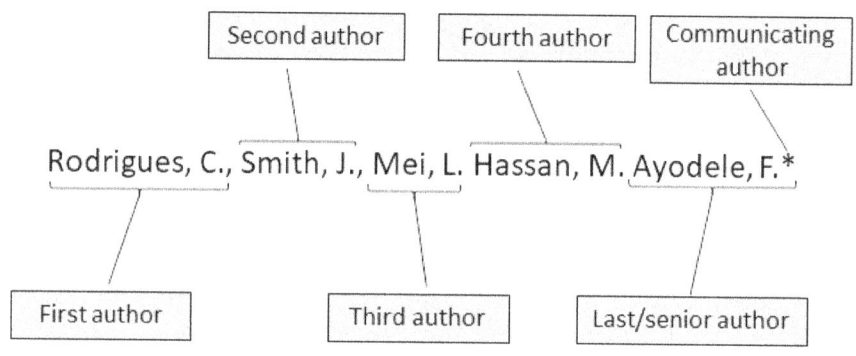

The *first author* is quite simply the first person in the author list. This person has done most of the work and has written most of the article. Sometimes this position can be shared by two people who have done the same work. In that case, there is usually a footnote with their names, saying something like "these authors contributed equally." Being the first author is important, as the article will be cited by your name forever. For instance, if your last name is "Chen," your paper will be cited forever as "Chen et al.." This helps to establish your name as an independent researcher at the start of your career.

Just as the first author has done most of the work, the second author has done the second-most work. And the third, the third-most, and so on. I was once the 26th author of a paper with 40 authors. You now understand that I contributed relatively little to that article!

The last position in the list of authors is usually reserved for the *senior author*. This is the person who oversaw the work. For articles written by MSc and Ph.D. students, this is usually their advisor. Or it is some big-name professor who runs the lab that produced the work. Everyone knows that this person has sometimes made few active contributions to the article, so there is no need to get upset about having to include this author in your author list. Everyone understands that the senior author's role can sometimes be limited to supervision, financing, or providing lab space. Nevertheless, it indicates the senior status of the author and can therefore be a desirable position in the author list for middle-career scientists.

The *communicating author* is the person who submitted the manuscript to the journal and who managed the whole submission and review process. The communicating author is not indicated by their position in the list, but their name is indicated with a symbol and is often the only person that has their contact information, such as their email address, listed on the article. This is the person to whom people should address their correspondence about this particular article. Usually, either the first author or the last (senior) author is also the communicating author. It shows who was most responsible for the article. If you are a student who just did what their advisor said, then probably your advisor should be the senior and communicating author, and you should be the first author. If you are an advanced Ph.D. student or postdoc, and you have conceived and developed the whole research yourself, you should probably be both the first and communicating author. Of course, there are differences between scientific fields and institutes in how this is done.

Many journals now also require a section called "Author Contributions," in which you indicate who did what. That section gives a lot more details on what each author did than the order of the author list.

THE ABSTRACT

After the title, the abstract is the most-read part of the article. It should summarize the article's major points, including the research question or hypothesis and methods, and discuss the major results. Just like the title, this should be a good reflection of the article's content and help readers decide whether they should read the rest of the article. Some journals ask that the abstract be organized into different parts, like an introduction, results, and a discussion part. Even if your article is behind a paywall, or otherwise hard to get, the abstract is usually indexed by search services, and readable to the whole world. Most of your readers will not get past the abstract, so it is crucial to make it count! For that reason, and because the number of words or characters is usually limited, authors spend much time getting the wording of the abstract just right.

KEY WORDS

Most journals ask you to give some keywords, so the article can be appropriately indexed. This is a holdover from the days before whole articles could be automatically indexed. When I was an MSc student in the 90s, we could only search the literature by keywords on CD-ROMs that were periodically sent by snail mail to the university library. Although most search services like Google Scholar now instantly and automatically index the Abstract or even the entire article, the keywords section is where you can put terms and names of concepts that characterize your article, but that you may not have used in the text. So it is still an important section to ensure your article will be found by its proper audience. The keywords can also state what broader topic or subfield the article contributes to, even though those words have not been used explicitly in the text. For instance, if your article deals with the evolution of a certain organism but has not used the word "evolution" in the text, you can use it as a keyword in your article. The keywords should not contain words already in the journal name or the title of your article. If you submit your manuscript to the Journal of Animal Ecology, for example, you don't need to use "animal" or "ecology" as keywords. Have a look at some papers in your field to see what the most common keywords are.

2B. THE BODY OF THE ARTICLE

INTRODUCTION

The introduction is there to place the research in the broader scientific context. It usually funnels the reader from a broad scientific issue to the specific question addressed in the article. Along the way, the reader is given all the necessary background information to understand the issue and why addressing it is important.

It can often be divided into roughly the following parts:

- A broader issue in the field that is not resolved and why it is important to do so
- The topic is taken from the broader issue to a specific or exemplary case
- The author gives the context you need to understand that specific case or issue

- Finally, the specific hypothesis/es to be tested is/are clearly stated. The authors often state what they expect the results to look like based on the hypotheses. E.g., "If our hypothesis is false, we expect our experiment to yield...".

METHODS

The next section is the Methods section, sometimes known as Materials and Methods. This part is very straightforward and explains how the study or experiment was carried out. It needs to contain all the details so that a researcher in the same field can duplicate the study. It describes the materials or samples, how and why they were selected, the experimental or observational procedures, the analysis methods, and the associated software. This section may not contain any results - that comes next!

RESULTS

This section contains a text description of the most important results and refers to any other representations of results, such as in images, tables, or supplementary files. This section can start by describing the dataset with some images or statistical descriptions. The results relative to the testing of the hypothesis should be well presented, but should be presented in a relatively neutral way, certainly without interpretation. Interpreting the results in light of the hypothesis and discussing their significance must be limited to the next section, the Discussion.

DISCUSSION

In the Discussion section, the authors refer to the results and say how they relate to the hypotheses tested. Was the hypothesis supported or falsified by the data? What can we conclude from the results? This is also the part within the manuscript where the authors can interpret their data and speculate about the implications of the results. The more conscientious authors will also warn their readers about the limitations of their study here. Sometimes a journal will also require the research conclusions to be in a separate section from the discussion. If that is not the case, the Discussion usually ends with the conclusions as the last paragraph.

After that comes only the literature list of the articles cited in the main text and the Acknowledgements section, where the authors thank all the people who contributed to the work but were not authors. Some journals also ask for other specific short sections, such as an Ethics Statement, the Author Contributions, where the data can be found, and so on.

3. BEFORE YOU START WRITING

THE METHOD

The method we will use here involves three steps. It is intended to write as quickly and efficiently as possible while staying close to your research outcomes.

1. First, we will figure out what <u>story</u> you will tell. You know many details about your study! Maybe you are among the top experts in the world by now. But from all that knowledge, you will need to carve out a coherent story that fits into the format of a research article.
2. Then, we will create and refine the <u>outline</u> for that story. By doing so, you will start thinking about what will be included in your article, and in what order.
3. Last, we will start <u>writing</u> sentences and paragraphs. With those two previous steps done, this part will be a breeze!

FIRST THINGS FIRST

Before we start writing your article the easy way, you need to figure out two things. You need to get an idea of what your story will be like and decide what journal you will send your finished manuscript to. These two things are interdependent- What kind of story you will tell about your research depends on the audience and scope of the journal you want to submit to. But to know what journal would fit your manuscript, you need to have a rough idea of the story you want to tell about your research. So we will first need to figure these two things out

3A. DECIDING ON A JOURNAL

Since you are doing the research or have already finished it, we will presume you know what your article will be about. But to which journal are you planning to submit it? It is important to decide which scientific journal you will submit your article to, as many journals have specific requirements for length, formatting, subsections, etc. If you know these in the early stages of writing, you can take them into account. Also, how you frame and tell your story will depend a little on the journal you intend to send it to; you want to write it so it is relevant to that journal's readership. So if you start writing before knowing what the journal wants, it may result in your work being a poor fit for the journal or require you to re-write large sections of your manuscript. And re-writing is a waste of your time, so let's decide on a journal in the early stages of writing.

But how do you pick a journal? That can be tricky if you don't know all the potential journals for which your article could fit. And since you are probably in the early stages of your career, you will not know all the journals well enough. Fortunately, there are a few options:

- You could use Google Scholar to find recent similar papers and see which journals they were published in (scholar.google.com). It is important to realize here that when a method is new, it can be published in journals with a higher impact factor than when it has been around for a couple of years. So if you are using a method that was "hot" five years ago, chances are you will not be able to publish in as highly valued journals as five years ago.

- In addition, you could use the online service Semantic Scholar to see which papers are connected to papers similar to yours (semanticscholar.org).
- You can look up a review paper on your field or subject, and look in the literature section to see which journals publish most articles in your field.
- If you already have your manuscript, or part of it, written up, you can use the website, Jane. Here, you can paste the text of your manuscript, and Jane will find similar papers (jane.biosemantics.org).
- The last and best tip is to ask your senior co-authors or colleagues. They have been in your scientific field for longer and will know which journals are a good fit for your research and the results you got. They can also give tips on framing your research to fit a particular journal.

But if you have not picked a journal yet, don't worry - you can already start writing the Methods and Results sections! You will learn why and how in the following chapters. But let's first figure out what you are going to write...

3B. SHAPE YOUR STORY

Beyond the jargon and fixed format, each research paper tells a story - Stories are how we humans organize our knowledge. Organizing your research as a story makes it easier for your readers to remember what your paper is about. So what story about your research will you tell the world? Your results say something about your hypothesis and have implications for the narrow topic of your research, but they also add a little brick to the wall of the much broader field in which your research is embedded. Having the story of your research clearly in your mind requires understanding what is essential, what is less important, and what it all means and implies. Nothing makes writing a paper more straightforward to write than having a clear story to tell. So how do we create a story?

Of course, *you* know what your research is about- You spent your blood, sweat, and tears on it! Since it is a scientific study, there are many meaningful connections to other works, and your results may answer several questions or tie into several subfields or issues. You probably know all these subtle angles and connections because you have spent much more time thinking about this problem than most people. And during your literature searches, you probably have gotten a pretty good idea of how your research fits into the field. But how can you tell this complex story to someone else? How do you shape the tangled spiderweb of your knowledge into a clean and easy-to-understand story without leaving out any important parts? You will need to invest a little time and thought to shape your story. This is the part that most people tend to skip, while it is the most essential. Most people think that they can tell the story right away because they know all the details. This is a mistake: knowing all the details, you may lose sight of the forest for all the trees. To guide your thoughts, here are a few questions as starting points to help you start thinking about your story. So stop reading and take your time to answer these questions for yourself. Saying the answers out loud, or writing them down, also helps. If there are any questions you cannot answer, you should do a little more background reading and/or talk to your advisor.

- <u>What was the problem you addressed?</u> What was the state of knowledge before you started your research? What knowledge hiatus or specific issue did you target in your research? What hypothesis/es are you testing specifically?
- <u>How did you go about trying to solve the problem?</u> You chose a specific method to address the problem. Why did you pick that one? What are its advantages and limitations?
- <u>What new data did you generate?</u> How are the results of your study different from other similar studies? How do your analyses compare to other studies?

- <u>What did the results show you?</u> Were you able to corroborate or falsify your hypotheses? Or did your results show you something you did not expect?
- <u>What implications do your results have?</u> Now that we know the results of your research, what does that mean for the issue you identified? Is it resolved, or is it redefined in some way? And how does that change the landscape in your field? What was left unanswered, and what new questions arose?

Did you answer each question in detail? So much deep thinking hurts the head! But now that you have taken the time to think through the main points of your story, we can start building the outline of your research article! Knowing the story enables you to structure your article and helps you decide which journal to send your manuscript to.

Now you are ready to start constructing the supporting skeleton of your manuscript, the outline!

4. CREATING THE OUTLINE

A research article has the existing structure of Intro, Methods, Results, and Discussion, but each of these sections has an internal structure too. You know your stuff, so if you start writing one of these sections, the story will come out, right? Wrong... Perhaps that is what it looks like when you read an article, but that is like trying to fly before you can crawl. Very experienced researchers can sometimes write that way, but that is only because they already have internalized the structure development for each section. They can organize the structure in their head- You are probably not there yet, so don't try. And science writing is complex; there are many different things you will need to introduce or reference to explain your research, particularly in the introduction. If you do not figure out the structure beforehand, you won't see the forest for the trees, and you will get stuck. Or you will have to re-organize it many times. Both are a waste of your time and are entirely avoidable. Making a logical story of a complex scientific problem is difficult, and most people spend much time re-structuring the text AFTER it has been written. This wastes time and makes for odd transitions, and it is easy to lose the overview among the enormous blocks of text. So that is why I found that making an outline of the structure beforehand, and filling it in, saves time and a lot of annoyance and confusion.

We will first create an outline, ensuring the story makes sense. Only then will we start writing. An outline is a backbone on which you will later hang all the text of your article. By creating the structure of your article in an outline, you will save yourself the re-arranging and re-writing of big blocks of text later. Also, this will ensure that everything has a logical flow and that all important bits are covered. Making an outline is great for any writing- I made a detailed outline before writing this book!

Some people first write the abstract and use it as an outline for writing the paper. Although this seems clever, the function of the abstract is different, and it is also quite short. It is too short to help you figure out all the essential details and flow of the story, so I recommend writing a real outline.

Let's create a text/Word file and start outlining your article. In your new text file, put all the titles for the Abstract, Introduction, Methods, Results, and Discussion sections. Now, for each of these sections, we will put a very rough structure in place that we will refine later. I recommend doing this using bullet point lists. Each bullet will later become a paragraph or more. Keep your bullets short to keep an overview of your article's structure. Long sentences at each bullet point make it harder to keep an overview. Structuring your article like this will help you get the order of the main story of your research right. If you have several things to say for each point, then just make several sub-bullet points. You do that by pressing the tab key after making a new bullet point in MS Word. Maybe not everything involved in your particular kind of research completely fits in this structure, but if your research needs more sections, you can add them as needed.

METHODS

The easiest part to write is usually the Methods section, so always start there. Typically, a Methods section is structured in roughly the chronological order in which each step of the research is done. So first, explain how you decided or calculated what subjects or materials you needed. Then what you did with them, your treatment, or how you did your observations or simulations. Then how you analyzed the data you got from your experiments or observations. So generally, the Methods section can be divided into these three parts:

- Materials
- Experiments (observations, simulations)
- Analyses

If your work involves multiple experiments or data gathering types, explain each in order. You can even make subheadings for different types of experiments, simulations or observations you made and explain the materials, experiments, and analysis steps for each.

Possible structure of a Materials section:

- Materials
  - Choice of your test system
  - Choice of sample size/measurement method/etc.
- Experiments (observations, simulations)
  - Experiment 1
  - Experiment 2
- Analyses
  - Data treatment 1
  - Data treatment 2
  - Data analysis step 1
  - Data analysis step 2

Another possible structure of a Materials section:

- Materials system 1
  - Choice of test system 1
  - Choice of sample 1 size/measurement method/etc.
- Experiments system 1
- Analysis results system 1
- Materials system 2
  - Choice of test system 2
  - Choice of sample 2 size/measurement method/etc.
- Analysis results system 2
- Combined analysis systems 1 and 2

What you put in your outline depends, of course, on what kind of study you did. Here are the starts of example outlines of Methods sections from two different fields.

Biophysics:

- Preparation of lysosomes
- X-ray diffraction
- Neutron spin-echo spectroscopy
- Molecular dynamics simulations
- Dynamic light scattering
- Lipidomics analysis

Based on: 10.1371/journal.pone.0269619

Atmospheric circulation:

- Coring, sub-sampling, and geochemical measurements
- Composite profile and radiocarbon dating
- Plant macrofossils
- Stable isotopes

Based on: 10.1371/journal.pone.0277027

RESULTS

For this section, start a bullet list and write down "describe dataset" as the first point. Then, you write down bullets for each type of result you have in the order that they appear in the Methods section. You probably end up with 3-6 bullets in total. Ensure that the most important results, those that deal with testing your hypotheses and on which your story depends, are in there!

Example structure, following the order of the Methods section:

- Brief description of your dataset.
  - Overall statistics
  - interesting observations, etc.
- Results of data analysis step 1 (in words and/or by referring to tables/figures)
- Results of data analysis step 2 (in words and/or by referring to tables/figures)
- Etc.

Remember- objective results only, no interpretation of your results in this section!

Prioritize presenting results that are most relevant to your hypotheses, e.g., by making nice images or tables.

Molecular phylogenetics:

- Describe dataset
  - Number of samples/species
  - Number of genes sequenced
  - Size of final alignment
  - Error rates, quality scores
- Result model testing and selection
- Result phylogenetic reconstruction
  - Describe overall tree
  - Refer to Image 2: phylogeny
- Result trait mapping
  - Describe trait distribution
  - Refer to image 3: Trait map
  - Statistical testing
    - Trait correlation

Clinical medicine:

- Describe dataset
  - Demography patients (table)
  - Treatment history
  - Symptom severeness
- Symptoms and impact
  - Interview group composition
  - Symptom prevalence (table)
- Treatment priorities
  - Patient-tailored treatments
  - Best treatment outcomes
    - Scenario B
    - Scenario C
    - Scenarios A and D
  - Endpoint preference
    - TTD (figure)

Based on: 10.1371/journal.pone.0280259

DISCUSSION

This section is less standardized in outline than the Methods and Results sections. To start, try to make a bullet point for every suggestion below. It doesn't have to be perfect yet, but the order has to make a logical story. As this section tends to be less standardized, your story may not fit the structure I suggest below. If that is the case, make your own logical order. All that matters is that you convey the story you want to tell so that others can easily understand it. Referring to the last two story-defining questions in the previous chapter may help you with that.

- Briefly remind the reader of the issue/hypothesis being addressed.
- Your interpretation of the main result(s) of the work
  - For each result in the Results section, explain *what* you found (without repeating the results) and *what you think it means* in light of the hypothesis you were testing and the specific issue from the Introduction section.
- Discuss any interesting and unexpected observations or patterns you encountered.
- Discuss the scope and limitations of your study.
- Discuss how your results and their interpretation change the knowledge landscape in the field or pertain to the broader issue.
- Discuss possible next steps to be taken to further address the (broader and narrower) issue based on your results.

INTRODUCTION

Now that you have a rough idea of what you want in your Discussion section, you know what to introduce in your Introduction section. These two parts depend on each other: The hypothesis you made in the Introduction gets its final verdict in the Discussion. The problem introduced in the Introduction will be discussed, and perhaps resolved, in the Discussion section. Your work may impact the state-of-the-art in your field, so that will need to be introduced in the introduction. You will find yourself going back and forth when writing the structure of these two sections. When writing the outline of the Introduction section, you need to put many sub-bullets in the background section. Again, it does not need to be perfect, but try to make the order of your story's steps logical.

- The broad scientific issue, problem, or hiatus in knowledge, its background, and why it is important.
- Funnel from the broad scientific issue to the particular issue that will be addressed in your article; explain why this specific issue is a good/important example for the broad issue.
- Give all the background information necessary for a scientifically literate general reader to understand the issue you are addressing. Depending on the issue, this can be a large part of the introduction containing many citations. When describing the state-of-the-art, be careful not to write an entire review of the field. Only the papers that are directly relevant to your research should be cited, and you can give a reference to a review paper for people who need an overview of the field.
- Formulate how the issue can be addressed by testing a specific hypothesis. State the hypothesis/es you will test explicitly. Many journals now prefer hypothesis-driven papers, so this is vital.
- Briefly explain what you expect to find based on the state-of-the-art and the hypothesis. There is a trend among some authors to hint at how it is done and the results, although that kind of information does not belong in the introduction section. It is unnecessary to create suspense in this way: the reader will already know what is coming from reading the abstract

These are all the sections that you need to outline at this stage. I recommend reviewing the whole thing several times, taking a break, and returning to it later to review and refine it with fresh eyes.

REFINING YOUR OUTLINE

You should now have a rough outline of the most important parts that will make up each section. Go back to it, and see if it still makes sense to you. Is the order correct? Can you imagine how you will make a logical transition between each part? If not, change the order, or add more points. You are now already refining your article outline. But you can make the writing even more effortless by doing another pass over your outline and adding another level. So before we start turning your bullet points into paragraphs, we can subdivide them into sub-bullets. That will only be necessary for some of them, but it will help with the long paragraphs, such as those in the background in the Introduction section. So let's go over each of your bullet points and try to make sub-bullets where possible.

For instance, in the results section, you could subdivide your first bullet, the one where you describe your dataset, into each aspect of the dataset you want to mention. For instance, you may want to say how many data points you have, what the minimum, maximum, and mean were, if there were any outliers, and so on. You see that each sub-bullet will eventually become one or a few sentences. By adding them as sub-bullets and not yet writing them out completely, you can easily keep an overview and make sure everything is in the order you want. In the Results section, you can, for instance, also have the WHY and HOW for each step as a separate bullet point (see next chapter for more on the Why and How of the Results section).

The example structures in the Discussion and Introduction sections sometimes already have sub-bullet points. Now take your time to review your entire structure and subdivide each bullet point as much as possible. This will make writing the text later on a breeze! Don't leave anything out that you think should be in the article.

Here is an example of what the previous Methods section outlines may look like when (partly) refined with sub-bullets.

BIOPHYSICS:

- Preparation of lysosomes
  - Collection of blood samples
  - Prepare RBC liposomes from samples (cite protocol)
  - Creation of RBC ghosts through washing and centrifugation
  - Creation of LUV solution (tip sonication)
  - Apply to silicon wafers for X-ray experiments
  - Neutron spin-echo experiments
  - Measure vesicle diameter via DLS prior shipment.
- X-ray diffraction
  - Describe instrument and setup
  - Etc...
- Neutron spin-echo spectroscopy
  - etc...
- Molecular dynamics simulations
- Dynamic light scattering
- Lipidomics analysis

Based on:
https://doi.org/10.1371/journal.pone.0269619

ATMOSPHERIC CIRCULATION:

- Coring, sub-sampling, and geochemical measurements
  - Describe core
  - Cutting core
    - Half cellulose extraction
    - Half Isotope analysis
  - Fluoroscope experiment descr.
- Composite profile and radiocarbon dating
  - Sample selection/discarding
  - C14 analysis descr.
  - Age-depth model
- Plant macrofossils
  - Sampling method
    - Interval
    - Sieving
    - Reference materials
- Stable isotopes
  - ...

Based on:
https://doi.org/10.1371/journal.pone.0277027

READY TO WRITE?

So, are you happy with the structure of your story? Are you sure you captured everything you want to say? If you have co-authors, consider sitting down with them to discuss the order and structure and whether anything is missing. That way, they can check if anything needs to be added. Co-authors can sometimes ask to re-order entire sections, which will take a lot of time. Better to address and discuss this in the outline phase and decide on the final structure together! That way, they cannot ask you to re-write or re-order the whole thing later.

Once your structure is complete, copy the entire thing. Now you should have two copies of your structure in the text file: one to keep as an overview and one for you to fill in! You can replace each sub-bullet with sentences now. And if you get lost, you still have the first copy of your bullet-list structure to refer back to. Since you now know what goes where, you should have no problem writing a paragraph or two for each bullet point. Put them all together, and you should have most of your article written! And that is all there is to this method. You are ready to write your article.

But what about the Abstract and the Title? You write those after you have written the rest of the manuscript. When writing, you will get new ideas, change the story a little, and get an excellent feel of the central message of your article. Once that is all settled, you can write your title and abstract. We will discuss how to write those sections once the rest has been written.

Remember, it is best to select the journal at this stage, when you have your story figured out but have yet to write the article. The story determines what journal is the best fit for your article. If you have terrific, broad-reaching results, you can get into a broader or better journal than if your results have limited or specialized implications. Once you know which journal you are going for, you can study the Guide for Authors for that journal to see what they require. You can also look at some similar articles in the same journal. That way, you can be sure you will fit the text of your article into a format, style, and tone that the journal wants, also fitting the scope and audience of that journal. We will start writing out the article's full text in the next chapter.

5. SCIENTIFIC WRITING STYLE

Now that you have your detailed outline of each main section ready, we can start replacing each bullet point or sub-bullet point with a paragraph of text. But before you start doing that, there are a few stylistic rules to keep in mind:

- Science writing is about clear communication and accuracy. Science is also an international, multicultural enterprise, so avoid vague or ambiguous terminology and be to the point. E.g., instead of saying, "we attempted to execute," just write, "we tried." >><u>Use short, clear sentences in simple but accurate language!</u><< (if I could put flashing neon lights around that sentence, I would)
- Give specific quantitative values when possible. So instead of "a high pressure" or "highly significant," write "a high pressure of 10.000 GPa" or "at a highly significant $p<0.0005$".
- Don't present small sample sizes as percentages- Write "four out of the six patients recovered" instead of "66.67% of patients recovered".
- Whole numbers zero to ten (sometimes up to twelve) are spelled out. E.g. "Out of 22, nine participants were selected..".
- Writing in either the first person singular or plural ("I did," "we studied") or the third person ("was done," "was studied") depends on the journal and the number of authors in your article. Check some recent articles and the Guide for Authors of the journal you intend to submit your article to. Once you have picked a person to write in, be consistent.
- Tenses:
  - Use the Present tense to describe currently accepted knowledge, e.g., "Mars <u>is</u> further from the sun than the earth."
  - Describe previous studies in Past Imperfect tense, e.g., "Wu et al. <u>showed</u> that radiation reduces...".

- o   The Present Perfect tense can describe ongoing or uncompleted processes or things that have not yet been done. E.g., "Journals <u>have increased </u>the rate of..", "Recently, Richards et al. <u>have investigated </u>..".
- o   Use Past tense to describe your hypothesis and objectives, e.g., "We <u>hypothesized</u> that..", "We <u>aimed</u> to..".

With that in mind, let's start writing!

6 WRITING OUT EACH SECTION

6A. FIRST, WRITE THE METHODS AND RESULTS SECTIONS

METHODS

In your outline, you should have bullet points to describe the subject/population of the study, the preparation of your subjects or equipment, the protocol of the study, and how you measured and analyzed your outcomes. Now, you will write a detailed paragraph or two for each bullet point.

When writing your Methods section, follow this structure for each method or analysis step: First, *Why* and then *How*. For instance, you could start by explaining *Why* you would need to select specific study subjects and then explain *How* you achieved that selection. Then, you could go on and say *Why* you needed to perform a specific treatment and then *How* you did it. Then you go on saying *Why* you think it was necessary to remove the outliers from the final dataset and then *How* you selected them. And so on. This will help your readers immediately understand why you used those methods and also aid in logically connecting them to your research question and hypothesis. E.g., "To get a sample size sufficient to test both X and Y, we selected more than 100 ...." or "To linearize dimensional scaling, all measurements were log10 transformed before analysis..". Following this *Why* and *How* structure, your

Methods section will go from a dreary listing of seemingly random technical methods to a logical sequence of clever steps you performed to achieve your goal. Your Methods section will immediately read better than that of most articles out there.

A few specific notes for the Methods and Results sections:

- If you are using a protocol that is already described in another paper, there is no need to repeat it completely. Just put a reference to the paper that has the correct protocol. E.g., "DNA was extracted as described in Smith et al., 2007". Sometimes it is helpful for the reader if you write a short version and give a citation for the complete protocol.
- Always put species names in italics, and give the full name and authority when first mentioning a species in the article. E.g., "We measured five specimens of *Scorpio maurus* Linnaeus 1758. These *S. maurus* were.. ".
- Use the conventions of the IUPAC-IUB for chemical nomenclature.
- Use SI units for reporting any measurements.
- There are several reporting guidelines for specific types of papers:
  - CONSORT for reporting randomized controlled trials (The CONSORT statement: Revised recommendations for improving the quality of reports of parallel-group randomized controlled trials. Ann Intern Med. 2001;134:657-662)
  - STARD for diagnostic accuracy studies (Bossuyt PMReitsma JBBruns DEet al. Towards complete and accurate reporting of studies of diagnostic accuracy: The STARD Initiative. Ann Int Med. 2003;138:40-44)
  - PRISMA for meta-analyses (Moher DLiberati ATetzlaff JAltman DG. The PRISMA Group (2009). Preferred reporting items for systematic reviews and meta-analyses: The PRISMA statement. PLoS Med 6(6):).)

You will find that the details you put in the Methods section often differ from those in your lab journal. You will usually not write the name of the supplier of your materials or the software version in your lab journal, but these details are necessary for the Methods section. So that is why I recommend you start writing the Methods section while you are still doing your research and analysis, even when you keep a meticulous lab journal (If not now, maybe for your next article!). Writing down all the details while doing it is easier than figuring them out afterward!

RESULTS

And? Did your results confirm your hypotheses or not? Either way, you'll want to write them down so your readers can navigate them easily in the Results section.

The first and easiest thing to do is to describe the dataset that resulted from your experiment or observations—the number of data points and some interesting statistics such as means, minima, and maxima. Just describe the aspects of your dataset that may interest your readers. Of course, you will provide the entire dataset publicly somewhere (See the section on data availability), but why not give some interesting statistics or observations in your paper. About one short paragraph should do it. Sometimes there are noteworthy personal observations that are hard to capture in numbers. If you think these observations may be valuable to other researchers, you can describe them here. E.g., "The monkeys initially seemed reluctant to put on the mind-reading helmets."

The most important results are those related to the hypotheses you were testing and the conclusions you intend to draw from them. You should present the results for each type of analysis you did or for each statistical test. These should be given in the same order as you introduced them in your Methods section. If you have multiple experiments to present, give their results in the same order as you had them in the Methods section. Depending on the journal guidelines, you can make corresponding (named or numbered) subheadings for each experiment in both Methods and Results.

Some results are best presented as images, graphs, or tables. It may be worth making a picture or table for anything that cannot be described in two or three sentences. If you made any of these, you should refer to them here. It is best to do this in context, e.g., "The highest values were recorded at 3.21GeV (see Figure 3).". However, a short sentence like "The results of the flux capacitor time-dilation test are presented in Table 2." can sometimes already be enough.

When writing out the paragraphs for each bullet point in your outline of the Results section, remember that this section should not contain any interpretation of the results or other value statements.

6B. NEXT, WRITE THE DISCUSSION AND INTRODUCTION SECTIONS

DISCUSSION

In the Discussion section, you refer to your results and interpret them in the light of your hypotheses. Don't repeat the Results section or introduce new results here.

Most researchers are fine with writing the discussion. After all, the implications of their research are what they are most excited about and why they did it in the first place. But it is easy to go overboard here and oversell your results. To avoid overstating the importance of your results, you can soften your statements by using "this suggests," "possibly," "seems to," etc.

Having both interpretation of results and conjecture about their broader meaning in the Discussion section often leads to confusion. This sometimes leads to grandiose headlines and misunderstandings of scientific studies in the media. Make sure it is apparent in the Discussion section what follows from your result and what is pure conjecture. Both are allowed and even expected, but make sure your reader knows which is which. So make sure you make clear in your discussion section which parts are the corroboration or falsification of your hypotheses and which parts are merely your (presumably well-informed) conjectures about their meaning and importance. You can do that by starting your Discussion section with one or more paragraphs about the interpretation of your results in the light of your hypothesis. In later paragraphs, you can then start talking about what the corroboration or falsification of your hypothesis means for the greater scope and what it may mean for future research and impact.

Something we don't see enough in many papers is a review of the scope and limitations of the study. This is because some researchers feel they should only highlight their research's strengths and hide the limitations. This results in overselling the results in the paper itself, in papers citing it, and in the media. I don't recommend hiding the limitations of your research. In fact, I am much more confident in research articles in which the authors seem aware of both the strengths AND the weaknesses of their research. It shows a greater awareness and responsibility toward the reader. Knowing the limitations of your study is a strength, so don't be shy about discussing them.

INTRODUCTION

What you choose as the broad scientific issue is sometimes more art than science. People often try to shoehorn their particular research into a fashionable broad topic that has little to do with it. But not every research article has to be about climate change or a cure for cancer to be a valuable addition to human knowledge.

When giving the background information necessary to understand the specific issue you address in your research article, it can be hard to know where to draw the line. Some reviewers will want you to make a mini-review of the entire field, while others will tell you that you wrote too much. Don't make your introduction a mini-review. As a rule, only give the background information necessary and sufficient for a scientist in your discipline to understand the problem you are addressing with your research. Anything more may confuse the reader. The reader should be able to understand the problem without reading the referenced articles. If you are unsure, ask a fellow physicist, biologist, or astronomer who does not know your research field to read the introduction. They should be able to understand the problem you are addressing from just reading your Introduction. If you want, you can help readers who don't know your field's background by giving some references to review papers that provide more background information.

6C WRITING THE REMAINING SECTIONS

THE ABSTRACT

You will find that while writing, your story can still evolve a little. Perhaps you change the relative importance of one result over the other, or you found a new reference that slightly changes how you interpret your data. That's why I recommend waiting to write your Title and your Abstract to the very end. The Abstract is just a very condensed version of the entire article, and now that you have just written the whole article, it will be straightforward for you to write the Abstract. In short, write what was done and the main findings. If you have space left, you can add a little conclusion at the end. For example, your abstract could look like this:

- Problem statement
- Hypothesis/es
- Main method(s)
- Main result + Discussion of their meaning/implications
- Brief indication of other interesting results (space permitting)
- Brief conclusion (space permitting)

All journals impose a limit on the number of words or characters in the Abstract. I find it best to first write everything that I think needs to be in there, and only then start to reduce the number of words or characters by cutting out unnecessary words. Also, references, acronyms, and abbreviations are not allowed in the abstract.

THE TITLE

And now that you have condensed your story into an abstract, you can condense it even more into a title. The title describes the content of your paper in the fewest possible words. So don't waste words like "A study of.." or "Observations on..". Contrary to the popular press, scientific articles' titles should honestly reflect what is in the article. It is a matter of scientific integrity to avoid making click-bait titles. Please help your potential readers by making a descriptive title. That way, people know what to expect in your article. But once you have put the essence of your article in the title, it is fine to make the title more interesting or fun. Clever wordplay helps get noticed and remembered, but that should never go at the cost of being clear, accurate, or descriptive. If you are at a complete loss or need inspiration, you could try to use a title generator such as writefull.com/title-generator or another AI language model, such as the GPT series.

DATA AVAILABILITY

Most journals ask that you submit your data and code to an online repository, such as Dataverse, Data Dryad, GigaScience, or Mendeley Data. This is necessary to assess the reproducibility of your results and makes the results of your hard work available to future scientists. You then give the DOI of your data in this short section, e.g., "The data has been made available through DataDryad, (DOI: doi:10.xxxx/dryad.xxxx; datadryad.org)". If you have a relatively small dataset, you can submit the entire dataset as supplementary material, for instance, as a comma-separated text file. In that case, you can write something like: "The full dataset is made available as a supplementary file to this article.". It is increasingly important to publish your analysis code, e.g., R, Matlab, or Python code, with your article. You can submit it to any of the above data repositories or make it available through GitHub if it is functional for other researchers.

6D. FIGURES AND TABLES

"Blot, blot, Western baby, figure one will be amazing."

- Zheng Lab - Bad Project (Lady Gaga parody)

FIGURES

Figures are a vital part of any scientific research article. They help to illustrate the points that you are making in the text and can be used to support or refute your arguments. As such, it is crucial to ensure that your figures are correctly formatted and presented so that the reader will immediately grasp what you want to convey with the figure. A whole book could be devoted to selecting, formatting, and presenting visuals like images and graphs. You can find many resources to help you with creating stunning visuals. Here we will go over the most important things and most common mistakes.

- When choosing which <u>type</u> of figure to use, consider the type of data that you are presenting and what type of figure would best illustrate it. Common types of figures include graphs, plots, chromatograms, blots, maps, and photographs. If it can be described in two lines of text, then don't make a figure at all.

- <u>Label</u> your figures clearly with a title and any relevant labels, axes, and scale bar. This will help readers understand the figure more easily. Apart from the necessary indicators for axes or other parts of the figure, the use of text on a figure should be minimized. Consider the final print size on the page or PDF when deciding on font size, and optimize for readability. Images' font sizes should be similar to the font size of the main article text.

- <u>Size</u> your figures to fit within the margins of the page without being too small or too large. Keep in mind that articles are often published with two-column layouts, and if so, consider making figures to fit one or across two columns.

- <u>Color</u> can be used to enhance a figure but should be used sparingly and only when necessary. Too much color can make a figure difficult to read or interpret. Use colors in combination with symbols where possible, or check your colors for accessibility for colorblind people (e.g., whocanuse.com).

- <u>Captions</u> should accompany each figure and briefly describe what is being shown in the figure and any relevant information about how the data was collected or analyzed.

- Cite <u>Sources</u>: If you use images or data from another source, cite it in your caption or a separate note at the end of the article. This will help readers understand where the information came from and give credit to its source.

- Show your <u>uncertainty</u>. Use error bars or similar visual devices to show how confident the reader can be in the precision of the value you show in the image.

Graphs can be produced in R, SPSS, Excel, etc., and further enhanced using, e.g., Illustrator, Inkscape, or another vector editing software. Photos and drawings can be edited in

Photoshop or Gimp image editing software, but each journal usually has some rules on how much a picture may be enhanced. More customized illustrations, such as composites, flow charts, or schematics, can be created or edited in Illustrator, Inkscape, or Powerpoint. For biology, Biorender.com is a great online tool for making simple schematics.

Here is a common mistake in data presentation. Sometimes you can see written under an image that "the x-axis means this, and the y-axis means that." Although it is traditional to plot the x variable on the horizontal axis and the y variable on the vertical axis, that is not the correct way to indicate these axes if you are plotting variables with other names on them. Just refer to them as the horizontal and vertical axes, e.g., "..mass is plotted on the vertical axis..".

IMAGE CAPTIONS

When you write captions for tables or graphs, tell your readers what points they should take away from them. After all, you made the image or table to show something in particular. Of course, you should write all the details necessary to understand the table or graph, but then also point your readers to what you think they should notice. Starting a caption like "Figure 1. The decline in resonance over time.." can be helpful to focus the attention of the reader on the salient pattern. Or, you can put a short sentence at the end of the caption, such as "Note the C-shaped curve in the arc in the lower left-hand side of the image." And, of course, ALWAYS label all your axes, and give the units of the measurements you show in all tables and figures.

TABLES

How your table will be formatted depends greatly on what you will put in it and what you want your readers to take away from it. As with images and graphs, making it easy for the reader to see what you want them to see is important.

Here are some pointers for formatting tables:

- Ensure the table is clearly labeled and numbered.
- Use a consistent font size and style throughout the table. Consider how large the table will be in print/PDF, and if it will still be readable at that size.
- Use column and row headings to clearly describe the data in the table. Include the briefest possible description of what the column/row represents and any units, if applicable. If heading descriptors or table entries become too long, consider Including a legend or key explaining any abbreviations or symbols used in the table.
- Use horizontal lines to separate larger sections of the table, but not to separate individual rows or columns. Avoid vertical lines, grids, or shading.
- Include a brief description of the table in the text of the article, including any relevant statistical tests used to analyze the data in the table.
- Include all relevant information in the table, such as sample size, mean, standard deviation, etc., as appropriate for the study design and data analysis methods.
- Make it easy to compare values across rows or columns.
- Tables of more than one page should be moved to supplementary materials. These supplementary tables should also be referenced in the text.

A poorly formatted table. Descriptions are unclear, units are missing or not in SI units, and the number formatting makes it hard to compare the values across cells. The grid ruling gives a messy and cut-and-paste impression.

Spec	n	#	Speed	A (m/s^2)	Path (inch)
A. amoreuxi	5	51	96.1	8.18 ± 2.79	3.515 ± 0.94
A. australis	3	30	110.4896	9.25956 ± 2.97	8.97 ± 2.59
H. franzwerneri	2	12	90.89832	6 ± 2	11.068 ± 2.33
H. gentili	2	12	89.8087	5.82 ± 1.51	11.7 ± 4.5
Pa. transvaalicus	1	11	66.6195	5.59 ± 2.03	7.58 ± 3.56

The same data in a well-formatted table, showing the names and units of all variables and the standard deviation. Note that only two horizontal lines are used. Note also that non-integer values are given to three digits of precision (See chapter 7 for more about digits of precision). The column headers can be further described in the table caption, or the reader can be pointed to the relevant part of the text that explains each variable in more detail.

Species	n	# Strikes	Speed (cm/s)	Acceleration (m/s^2)	Path Length (cm)
A. amoreuxi	5	51	96.1 ± 19.1	8.18 ± 2.79	8.93 ± 2.43
A. australis	3	30	110 ± 27.4	9.25 ± 2.97	8.97 ± 2.59
H. franzwerneri	2	12	90.8 ± 12.7	6.88 ± 1.71	11.0 ± 2.33
H. gentili	2	12	89.8 ± 14.8	5.82 ± 1.51	11.7 ± 4.50
P. transvaalicus	1	11	66.6 ± 28.9	5.59 ± 2.03	7.58 ± 3.56

6E. REFERENCES

In the reference list, you should have all the references mentioned in the text. They should be formatted to fit the journal's preferred reference style. Most reference management software will automatically populate and update this list. You just need to tell the software where you want it. If your reference software does not have the correct style for your journal already loaded, you can usually download a reference style file from the web and implement it in the software. You can also create your own reference style files. Always check if your reference list is in the proper format for the journal, and that all articles that are referenced in the text have a corresponding entry in the list. With a good reference manager, keeping track of and formatting your references is automatic, so you can concentrate on writing rather than formatting.

You choose which papers you cite in the text. You should always give credit for others' ideas where it is due. But on which giants' shoulders are you standing? In other words- which papers should you cite in your work? Should you always reference the first, most foundational paper mentioning a particular fact, or rather reference only the latest, cutting-edge papers? That depends on whether your work builds upon or challenges a foundational idea or tags new data onto the latest state of the art. In the first case, you cite the old foundational papers. In the second, you mention the latest papers that you build upon. When doing so, it is best to not directly criticize other works, but state their objective strengths and weaknesses.

A much-discussed stylistic issue is whether references should be near the words describing the information you took from that reference, or should all references be at the end of the sentence. For instance, compare the following:

> *Recently, several studies have highlighted the need for this technique in research (Abate et al., 2019), clinical work (Bhatia et al., 2020), and clinical applications (Cavalcanti, 2021).*

with

> *Recently, several studies have highlighted the need for this technique in research, clinical work, and clinical applications (Abate et al., 2019, Bhatia et al., 2020, Cavalcanti, 2021).*

The first way of organizing the references clarifies which reference goes with which application, but the second makes the sentence much easier to read. So there are good arguments for both sides of this heated argument! You can avoid this issue most of the time by keeping your sentences short and simple.

And if you cannot avoid it, a rule of thumb is to place all references at the end of the sentence to avoid breaking up the flow of the sentence *unless* it can lead to a misunderstanding of what part of the sentence goes with which reference. So in the above example, it is best to keep each reference separate, as in the first sentence.

A similar example where all references could go together at the end is:

> *Recently, several studies have shown that climate model variable ratio correction is a valuable addition to the modeling toolkit (Abebe and Balogun, 2018, Erikson, 2020, Crespo and Delgado, 2021).*

Note that in such listings, the references should be in chronological order, with the oldest references first. If you have several references from the same year, you can put those in alphabetic order within that year.

7. COMMON MISTAKES TO AVOID

When helping students and editing or reviewing articles, I found that many people make the same mistakes when writing a scientific research article. Here is a rundown of some common mistakes. Make sure you don't fall into these traps!

1. STARTING TO WRITE TOO LATE.

Most people think they should start writing the manuscript after doing all the research. These people often end up struggling to get all the details of what they did for the Methods section. What was the measurement precision of my lab equipment? What brand of reagents did I use? Sometimes, they are even in a different town or country by the time they start writing, and figuring these details out will take more time than necessary. By starting to write the Methods section *while* doing the research, you can write down all the details as you go, when you have them available. Also, what you did is fresh in your memory - you will be surprised how quickly and easily you forget details that you currently deal with on a daily basis. So open a blank Word document when you start a new research project and start writing the Methods section as you are doing it!

2. STARTING WITH THE ABSTRACT OR THE INTRODUCTION.

Hey, since you start reading from the Abstract, why not start by writing it? Wrong. Your story will likely evolve while you are writing the paper. If you start with the Abstract, I will bet you have to rewrite it completely. I have seen it many times, so don't waste that time.

Likewise, starting the Introduction seems a logical place to start. After all, that is where the paper really starts, right? But as you have seen, the Introduction is tied to the Discussion section. It is best to start with the Methods and Results sections. If you have time and want to start working on the introduction before your research is over, you can - by searching for all the relevant literature and summarizing the relevant points of each paper. This will be a great help in writing the Introduction section later!

3. WRITING UNCLEARLY

Here is a common mistake I see in people whose first language is not English. They try to use fancy words and flowery language but quite often end up using the words wrong. If, like me, English is not your first language, then keep your language simple. Use the simplest and most precise words you can think of so that everyone can understand you. Native speakers, or people with more practice in writing in English, have a larger vocabulary and can select rare words that fit particularly well in a specific case. If you are less than fluent, don't try to emulate them. The goal of scientific writing is to write so everyone can unambiguously understand it, particularly those that may be less fluent in English than you. So keep your sentences short and your choice of words simple. Avoid convoluted, multi-line, many-comma sentences. That way, everyone can understand you.

For example, <u>don't</u> say:

> *"In the noble art of science writing, one must attempt their utmost to reduce unnecessarily complex, obscurant, pompous or ostentatious language and the associated*

incomprehensible verbiage and obscure vocabulary, and rather attempt to convey one's thoughts unambiguously using brief sentences of low complexity, constructed from words that have a relatively high frequency of usage in common parlance."

When you could say:

"In scientific writing, use simple and clear language." (If you were wondering why this book is so short, this is why 😊)

4. USING MULTIPLE WORDS FOR THE SAME CONCEPT

Another common stylistic mistake is avoiding using the same word or term twice or more in a row. Many people were taught at school to avoid repeating the same word in a text and to try to use a synonym. This is probably an improvement for prose and poetry, but it makes no sense in science writing. Since the goal is to be as unambiguous as possible, it is good to keep using the same word to refer to the same concept. Science is complex enough without using many different words to describe the same thing! Also, science terminology can be subtle, and what you think is a synonym for the same concept can actually refer to a slightly different concept. Remember, in science, terms have a precisely defined meaning that is often different from their meaning in everyday language. So in science writing, repeatedly using the same word for the same concept is preferable. It makes it clear that you are talking about the same thing. So remember: One concept, one word.

5. USING "RANDOM" WHEN YOU MEAN "ARBITRARY"

A widespread mistake is confusing the terms random and arbitrary. They may be roughly the same in everyday language, but they are clearly different in science. "Random" is a very specific mathematical property, whereas "arbitrary" means without preference. So if you selected your subject from a population "at random," that means you must have used some random number generator or random number set. If you did, you need to explain how you achieved this randomness in your Methods section. If you chose without a mathematical random selection mechanism but just arranged your subjects without preference, then you should use "arbitrary" instead.

6. NOT ROUNDING NUMBERS TO THE APPROPRIATE DECIMAL PLACES

OK, here is a big pet peeve of mine. People often copy and paste the numbers that a computer spat out when reporting numbers. You often see it in statistical indicators, such as p-values, reported to 6 or 8 digits of precision. The number of practical, informative digits is usually much lower. This shows that the writers do not understand what parts of the number are informative. If, for instance, a patient's body mass is 67.8936 kg on the scales, then the last few decimal digits are usually completely uninformative and should be rounded up. The body mass would change if the patient drank a sip of water, went to the bathroom, or even just breathed a few times. For most purposes reporting a body mass of 67.9 kg would be sufficiently precise. So think about what part of a number is informative for your specific case and what part is not, and round up appropriately. It will show you understand the relevance and value of the numbers you report.

So how does a scientist round up their numbers? An accountant, who always works with monetary values, can just round numbers off to the same number of decimal digits. This is what Excel does, for instance. You can set it to round to a specific number of decimal digits. But this is not useful if you are reporting different types of values. It is better to round things off to a specific number of digits of precision. This is what a scientist should do. Here are some examples of numbers rounded to three digits of precision:

3.97   891   0.000454   18.2   0.114   1.65e-10

It can be helpful to know that if you format your number to scientific notation, the number of digits of precision is the number of decimal digits plus one.

3.97e00   8.91e2   4.54e-4   1.82e1   1.14e-1   1.65e-10

7. NOT REPORTING UNCERTAINTY

Another thing to remember is that in science, you should always quantify how certain you are about your measurement or other results. That's why you should always report the standard deviation, confidence interval, or other measures of uncertainty with your numerical results and show error bars in your graphs. And when you see numbers reported like this [9.682041 ± 6.975768], you see how silly it is to report this many numbers of precision; The standard deviation is huge, so the measurement is imprecise, to begin with. So do your sensitivity analysis and report the measure of uncertainty -standard deviation, confidence interval, measurement error, or Akaike information criterion - whatever is relevant for your type of result.

8. HOW TO COLLABORATE

KEEPING TRACK OF VERSIONS OF THE MANUSCRIPT

You rarely write a research article on your own. Usually, several co-authors will write parts of the article or at least comment on them. So that means you will have several revised versions of your manuscript. Suppose one day you would continue writing in an older version by mistake. In that case, you risk losing important additions by your co-authors or even re-gaining errors that were already corrected. Therefore you must make sure you can keep track of the versions of your manuscript, so you can be sure that you always have the most recent version in front of you. Also, it can be helpful to refer back to older versions if necessary, for instance, to get back that paragraph you deleted. You can use version tracking software, but that is harder to do when you send your manuscript over email. People sometimes try to use version numbers to keep track of manuscript versions, but that will go wrong as soon as you have more than one author working on the manuscript: your version 4.1 may be their version 5! Version numbers tend to get confused very quickly during collaboration, and your file name may morph into something incomprehensible like: "Manuscript_ver3.2.4_edited3_Final2.1.docx".

A good option is to use Google Documents, an online document editing software. It is excellent at integrating additions by everyone, even at the same time. A drawback is that it does not yet work well with citation software. But if you have a moment when several people will work on a document simultaneously, for instance, during a conference call, Google Documents is your best friend. You can later download your document to be edited with a different text editor, for instance, to add the citations with your favorite citation manager.

I prefer to work asynchronously with my co-authors by sharing a Word document through something like Dropbox, Microsoft OneDrive, PCloud, or some other cloud drive. Or by sending them the manuscript through email. Either way, I always save a copy of the manuscript I am editing with my initials and today's date in the file name, e.g., "Manuscript_AvdM2-3-2023.docx". Every day I work on it, I will make a new copy with that day's date. So I will have a list of older versions available. By adding your initials and the date to the end of the file name, it is easy to see who last edited the manuscript, what version they started with, and when they did so, e.g., "Manuscript_PLC1-2-2023_AvdM5-2-2023.docx" is a file written by PLC on 1-2-23, and commented on by AvdM on 5-2-13. If you want to be especially neat, you can put the older versions in a separate folder. I put the date in the file name because sometimes the cloud software, or some backup software, can change the operating system time stamps on all the files. So don't depend on those automatic time stamps!

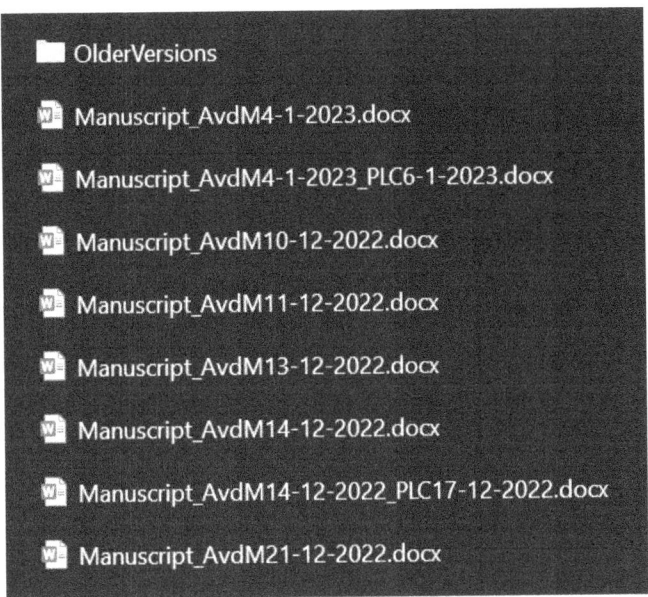

Figure 1 Screen capture of a folder with many versions of a manuscript. It is easy to follow the evolution of the manuscript when applying this naming scheme, even though they are not listed in order here. You can put older versions in a separate folder to remove unnecessary clutter.

COMMUNICATING WITH CO-AUTHORS

There are a few archetypes of co-authors: The busy hot-shot postdoc or young professor who has no time, the aloof big-name silverback professor, the unconfident technician, and the overly verbose MSc student. How can you make this odd collection of people collaborate efficiently? First of all, use a communication platform that all can use. I find even the most archaic emeritus dinosaur can use email. In the end, everyone needs to agree on the manuscript's version to be submitted, but you don't need to include everyone in every email. Sometimes it is quicker to deal with specific issues between two people without everyone else getting involved.

And how can you push that one unresponsive co-author along? Politely and by setting reasonable deadlines. You can say you will need their input by a specific date (I think anyone can find some time in two or three weeks). If no response comes by that time, give them a phone call, or in the worst case, move along while keeping them in the loop.

If there are disagreements between co-authors, it is good to listen to the arguments from both sides, as they will make valuable points for the Discussion section. Including the arguments from both sides in the Discussion may go a long way to appeasing them. Ultimately, it is up to the first author what gets incorporated into the manuscript. In all cases, your tone should be friendly, polite, and professional!

CITATION MANAGERS

Back when I did my Masters and Ph.D., there were no free citation managers, and the paid ones were buggy. So I had to format, sort, and edit all my citations by hand. When we decided to go for a different journal, I had to hand-format each citation in the list to the format desired by that journal. Now you can choose free citation managers to quickly organize and format your citations – Zotero, Mendeley, and Endnote, for instance. You can find comparison charts of citation managers on the web, so pick one used by your co-authors and enjoy.

9. THE SUBMISSION AND REVIEW PROCESS

TO PREPRINT OR NOT TO PREPRINT?

At this stage, you could consider publishing your manuscript on one of the many preprint servers, such as arXiv. Publishing a scientific research article on a preprint server has advantages and disadvantages. One of the main advantages of publishing on a preprint server is the ability to share research with the scientific community quickly and to receive early feedback on your work. Additionally, preprint servers can provide a more efficient and cost-effective alternative to traditional peer-reviewed journals. But preprint servers do not typically provide the same level of peer review and quality control as traditional journals. Also, search engines do not always index preprint servers, so they may not be as easily discoverable by other researchers.

SUBMISSION OF YOUR MANUSCRIPT TO A JOURNAL

Once you have finished your article, it is time to send it to the journal you selected. You should ensure the manuscript, figures, and tables are all in the format described in the "Guide for Authors" of that journal. And remember, each journal has different demands. Things to watch out for are figure and supplementary materials formatting; making mistakes here will cost you days, as the editor will spot mistakes and refuse your submission after some days. Then you have to go back and correct your mistakes and amend your submission before you can submit it again. This will delay publication.

Here are a few common things that can get your manuscript rejected:

- The research subjects, instruments, or methods are inadequate or flawed. An insufficient or biased sample is a particularly prevalent case.
- The statistical analysis is inappropriate, insufficient to test the hypothesis, or insufficiently described.
- The results are over-interpreted, and the study's results do not support the conclusions.
- The article is poorly written or organized.
- The research is not original or significant. Many journals receive many submissions and often reject articles that do not offer new insights or contribute to advancing knowledge in the field.
- The article is not relevant to the journal's focus or audience.
- The article does not meet the journal's formatting and style guidelines.

Some authors think they are clever by submitting a sloppy manuscript and letting the reviewers and editors tell them what details, like formatting of text, data, and images, need to be improved. Not only is this very disrespectful of the reviewers' time, but it will also cost you more time. The reviewers will need more time to review your paper, and you will have to go through more rounds of improvement and proofreading. Nothing is faster than writing a stylistically good paper, where the reviewers can concentrate on content rather than deciphering and improving your sloppy work. Remember - reviewers are your peers in your scientific field. They will see what poorly written manuscripts you tried to get away with, and they will adjust their opinion of you accordingly.

Even after going over your manuscript for the umpteenth time, there may still be mistakes that you don't see anymore.

Fortunately, there is plenty of help available, and a few are as follows:

- Use the spell check function in your word processing software to check your grammar and spelling. There are also excellent paid grammar-checking services, such as Grammarly.com.
- If you are not very practiced at writing in English, you can also enlist a proofreader or text editor through online service markets, such as fiverr.com or Upwork.com.
- You can check the formatting of your manuscript at Penelope.ai.
- Finally, to check the logic and completeness of your manuscript, consider having a more experienced colleague, such as a more senior co-author, read your manuscript before submission.

To submit an article, you usually have to make an account with the manuscript handling website of the journal. This can be through the Journal's proprietary online system, a centralized manuscript handling service such as "Manuscript Central," or sometimes through old-fashioned email. You will find the link to the submission system on the journal's website. Once you have followed the instructions to make an account, you can start a new manuscript submission. From there, just carefully follow the instructions. In particular, take the time to read the "Author Guidelines" for that journal. Submission can take some time, and I still set aside at least half a day to do it properly. In the submission process, you are often asked to suggest peer reviewers who can check your manuscript for flaws. Since you know who is important in your field, here is a great chance to put your article in front of them and ensure they read your masterwork! This increases the chances of the big shots citing your work.

Sometimes you can select an editor who is knowledgeable in your field during the submission process. The editor is the person at the journal responsible for dealing with your manuscript. The editor will select reviewers and ask them to review your manuscript. The editor will then go through the reviewers' recommendations and communicate them back to you. In the end, it is up to the editor whether your article gets accepted or not.

THE REVIEW PROCESS

If you made your submission correctly, you should see that your article is listed as "under review" in the online system. You will now have to wait for the peer reviewers to read and comment on your article. This will generally take at least three weeks but can often be up to six weeks or more. If it takes more than four weeks or so, you can contact the editor of your manuscript to ask about the status of the review process politely.

At some point, you will get a response back from the editor. The options are:

- Your manuscript is accepted as it is. This is extremely rare. You probably bribed someone or submitted to a predatory journal with very low standards. Be suspicious.
- Your manuscript is accepted with minor revisions. This means you have done a great job and need to tweak some things to make it acceptable for publication. Time to pat yourself and your co-authors on the back! Cake and drinks all around!
- Your manuscript is accepted with major revisions. This can be a mixed bag. Sometimes the major revisions that are asked are not so bad and take little effort to fix. Sometimes they require you to do the whole research pretty much again. In that latter case, the reviewers may be trying to politely reject your paper without giving you an outright rejection. Either way, you need to look at what the editor and reviewers commented and deal with each comment and suggestion, or decide to give up and try another journal.
- Your manuscript can also be rejected. If your work is not entirely flawed but just a poor fit, you can try another journal. If you did get comments from the reviewers, use these to improve your paper before you submit it elsewhere. Don't let the hard work of the reviewers go to waste!

When making revisions, it may sometimes be hard not to take the reviewers' comments as criticism. I recommend reading the comments one day, giving them some rest, and revisiting them one or more days later to make the revisions and write your replies. If you write replies to the reviewers' comments when you first read them, you may be a bit too hotheaded to reply politely and with the appropriate gratitude for the reviewers' efforts. In the end, the reviewers are often right, and you will learn to appreciate their candid and hopefully constructive appraisal of your work. If they gave you some really good pointers to improve your manuscript, consider thanking the reviewers in your Acknowledgements section.

When you go through each comment and suggestion, you should write whether you agreed with the suggestion and implemented it in the new version of the manuscript. Just create a separate file for your responses to the reviewers' comments where you put a reply after each reviewer's comment, e.g., "Changed in the manuscript as suggested." When a reviewer asks for clarification or additional information, don't just reply to the reviewer but add the missing information to the manuscript. The reviewers are usually not asking because they are curious but because they think something is missing in the manuscript! If you don't agree with a comment or suggestion, you can also reply to it and give good reasons why you think the reviewer misunderstood. You can disagree, but stay objective and polite! Remember, the reviewers are your peers in the field, and they will probably be for a long time!

There may be several revisions and reviews before your manuscript gets accepted. This can take months. It took almost a year for one of my earlier articles.

PROOFS

After acceptance of the final manuscript, including all the revisions, you will get the proofs from the formatting editor. Proofs are type-set versions of your article, ready for print and online publication. There are usually a few minor issues, called "queries," spotted by the formatting editor that need to be addressed. You usually need to reply to these proof queries and double-check the entire article, including captions, figures, and tables, within a few days. This is your last chance to make any minor corrections to the manuscript before it is published, so use a fine-tooth comb to find those last typos.

Before or after the proofs, you may also need to digitally sign some legal documents, e.g., about copyright. If you are publishing your article as Open Access, you often have to pay publication charges at this point. Some journals will give you a DOI number for your article as soon as it is accepted for publication so that you can include it in your CV, website, social media, or elsewhere. After that, you will be notified by the journal when your article is published. Congratulations! You are now a published scientist!

EPILOGUE

Thanks for reading this book on how to write a scientific research article! If you found it helpful and made your writing process easier, please let me know by leaving a review somewhere! If there are parts that were unclear to you, or if you have suggestions for additions or improvements, please get in touch with me, as that will help me improve the book in future editions.

For now, goodbye, and good luck with your scientific adventures!

DO'S AND DONT'S

Do:

- Start with writing the Methods section before you even finish the research
- Once you have your results, start writing the outline
- Use the experience of your senior co-authors and colleagues in selecting a journal before writing the paper
- Read the Guide for Authors of the journal
- Create an outline for your paper
- Use a WHY and HOW structure for the Methods section
- Use short sentences with clear, simple language
- Use the same word for the same concept throughout your manuscript
- Use a citation manager
- Put the date in the filename of each version
- Point the reader to salient points of figures and tables in the caption
- Label all axes, report units of measurement, and add scale bars to images
- Report the measure of uncertainty
- Make it clear in the Discussion what is supported by data, and what is conjecture
- Report both the strengths and limitations of your approach
- Make a descriptive title
- Set aside enough time for the submission process

Don't:

- Start with writing the Abstract or Introduction
- Make a click-bait title
- Try to shoehorn your research into a popular topic if it does not fit there
- Round to decimal digits, but use digits of precision instead
- Confuse the measurement with the phenomenon itself
- Submit a sloppy manuscript full of errors for the editors and reviewers to fix

ONLINE TOOLS AND SOFTWARE

Here are a few websites and software tools that can help you with the writing and submitting process. This is not an exhaustive list, just those that are mentioned in the text, and a few extras.

• Explainpaper.com	Ai-driven paper summarizer
• Writefull.com	Ai driven academic writing assistant
• Penelope.ai	Automated checking of manuscript formatting
• Endnote	Citation manager
• Mendeley	Citation manager
• Zotero.org	Citation manager
• Data Dryad	Data repository
• Dataverse	Data repository
• GigaScience	Data repository
• Mendeley Data	Data repository
• Grammarly.com	Grammar checking software
• biorender.com	Online schematic builder for biology
• Arxiv.org	Preprint server
• Whocanuse.com	Resource on colorblindness and how to design visuals for it
• Scholar.google.com	Search engine for scientific publications
• Jane.biosemantics.org	Search engine for similar scientific papers
• Semanticscholar.org	Search engine for similar scientific papers
• Fiverr.com	Work marketplace to find proofreaders or language editors
• Upwork.com	Work marketplace to find proofreaders or language editors

CHEATSHEET: THE STRUCTURE OF A SCIENTIFIC RESEARCH ARTICLE

Note: This structure is just a <u>suggestion</u> to help you think about how to organize your main story. This is the structure of a hypothesis-driven article; Your research type may require a different structure.

Although the sections are <u>presented</u> here in the usual order: Abstract, Introduction, Methods, Results, Discussion, Conclusions, they should be <u>written</u> in the order: (1) Methods, (2) Results, (3) Discussion + Introduction, (4) Abstract + Title.

TITLE (WRITE LAST!)

A clever play of words is nice for getting attention and being memorable, but <u>at least </u>make the title <u>a short and accurate description of what the reader can expect</u> to find in the article. Don't waste words here like "A study of...".

ABSTRACT

A condensed version of your paper. First, write everything you think needs to be in the Abstract, and only then start to reduce the number of words to fit the journal requirements. Some journals require subheadings within the Abstract, so check the Guide for Authors.

1. Problem statement
2. Hypothesis/es
3. Main method(s)
4. Main result + Discussion of their meaning/implications

5. Brief indication of other interesting results (space permitting)
6. Brief conclusion (space permitting)

INTRODUCTION

1. The broad scientific issue, problem, or hiatus in knowledge, and its background, and why it is important.
2. Funnel from the broad scientific issue to the very specific issue that will be addressed in your article; explain why this specific issue is a good or important example for the broad issue.
3. Give all the background information necessary for a scientifically literate general reader to understand the issue. Depending on the issue, this can be a large part of the introduction containing many citations, but don't write a review of the whole field.
4. Formulate how the issue can be addressed by testing a specific hypothesis. State the hypothesis/es you will test explicitly.
5. Briefly explain what you expect to find, based on the background information and hypothesis.

METHODS

For each method: first, explain WHY you needed to do it in this way and then HOW you did it.

e.g., "To get a sample size sufficient to test both X and Y, we selected ....", "In order to linearize dimensional scaling, all measurements were log10 transformed prior to analysis..".

Example structure, chronological order:

1. Choice of test system (Why and How)
2. Choice of sample size (Why and How)

3. Data treatment 1 (Why and How)
4. Data treatment 2 (Why and How)
5. Data analysis step 1 (Why and How)
6. Data analysis step 2 (Why and How)
7. Etc.

RESULTS

Remember - objective results only. No interpretation of your results in this section!

Prioritize the presentation of results that are most relevant to your hypothesis/es.

Example structure, following the order of the Methods section:

1. Start by briefly describing your dataset. Overall statistics, interesting observations, etc.
2. Results of data analysis step 1 (in words and/or by referring to tables/figures)
3. Results of data analysis step 2 (in words and/or by referring to tables/figures)
4. Etc.

DISCUSSION

1. Briefly remind the reader of the issue that was being addressed. E.g., "In this study, we tested the relationship between star luminescence and interstellar dust by testing (hypothesis)."
2. Give your interpretation of the work's main result(s) (those that pertain to the hypothesis/es you tested). E.g., "Despite the sound theoretical underpinnings, our results indicate no such relationship" or "Our results corroborate this hypothesis under these extreme circumstances."

  o <u>For each result</u>, explain *what* you found (without repeating the results) and *what you think it means* in light of the hypothesis you were testing and the specific issue from the Introduction section. If multiple interpretations are possible, discuss them all. E.g., "The positive correlation between variables 1 and 2 seem to indicate that ..." or "The large number of positive correspondents suggest this issue to be salient to the sampled population."

3. Discuss any interesting and unexpected observations or patterns you encountered, where possible, in light of the background from the Introduction section. If necessary, new background information can be introduced here to explain these unexpected outcomes.
4. Discuss the limitations of your study.
5. Discuss how your results and their interpretation change the knowledge landscape in the field, or pertaining to the broader issue.
6. Discuss possible next steps to be taken to further address the (broader and/or narrower) issue based on your results.

CONCLUSIONS

Depending on the journal, this can be a short separate section or a paragraph to end the discussion.

1. What your data allow you to conclude about the narrow problem you defined in the introduction
2. What your data allow you to conclude about the broader problem you defined in the introduction
3. Implications of this work for future research

Other sections include Author Contributions, Ethical Statement, Data Availability Statement, Highlights, etc.

• See the Journal's guide for authors.